MORAL FEELINGS, MORAL REALITY, AND MORAL PROGRESS

BY THE SAME AUTHOR

MORAL FEELINGS, MORAL REALITY, AND MORAL PROGRESS

Thomas Nagel

OXFORD
UNIVERSITY PRESS

OXFORD
UNIVERSITY PRESS

Oxford University Press is a department of the University of Oxford. It furthers
the University's objective of excellence in research, scholarship, and education
by publishing worldwide. Oxford is a registered trade mark of Oxford University
Press in the UK and certain other countries.

Published in the United States of America by Oxford University Press
198 Madison Avenue, New York, NY 10016, United States of America.

© Oxford University Press 2023

CIP data is on file at the Library of Congress
ISBN 978-0-19-769088-8

DOI: 10.1093/oso/9780197690888.001.0001

Printed and bound by CPI Group (UK) Ltd, Croydon, CR0 4YY

CONTENTS

PREFACE

This short book comprises two essays that are related to one another and that deal with questions that have occupied me for some time. The first, "Gut Feelings and Moral Knowledge," originated as a Dewey Lecture given at Harvard Law School in 2015. Toward the end of that lecture I made some remarks about moral progress. T. M. Scanlon was in the audience, and in the discussion afterward he posed a question about my invocation of that idea which made me think I had to say more about it. The result was "Moral Reality and Moral Progress," my contribution to a symposium marking the award of the Lauener Prize to Scanlon in 2016. I presented a greatly expanded treatment of the topic to the NYU Colloquium in Legal, Political, and Social Philosophy, conducted by Samuel Scheffler and Jeremy Waldron, in 2018; and I gave a version of it as the Jerusalem Lecture in Moral Philosophy at the Hebrew University in 2019. I am grateful for comments and criticisms received on all these occasions.

Both essays are concerned with moral epistemology and our means of access to moral truth; both are concerned with moral realism and with the resistance to subjectivist and reductionist accounts of morality; and both are concerned with the historical development of moral knowledge. The second essay also proposes an account of the historical development of moral truth, according to which it does not share the timelessness of scientific truth. This is because moral truth must be based on reasons that are accessible to the individuals to whom they apply, and such accessibility depends on historical developments. The result is that only some advances in moral knowledge are discoveries of what has been true all along.

The first essay was published in the *London Review of Books*, June 3, 2021.

The second essay has not been previously published, but the shorter Lauener Prize essay from which it derives was published in Markus Stepanians and Michael Frauchiger, eds., *Reason, Justification, and Contractualism: Themes from Scanlon* (de Gruyter, 2021).

T. N.
New York, October 2022

GUT FEELINGS AND MORAL KNOWLEDGE

1.

The philosopher Stuart Hampshire served in British military intelligence during the Second World War. When we were colleagues at Princeton he told me about the following incident, which must have taken place shortly after the Normandy landings. The French Resistance had captured an important collaborator, who was thought to have information that would be useful to the Allies. Hampshire was sent to interrogate him. When he arrived, the head of the Resistance unit told Hampshire he could question the man, but that when he was through they were going to shoot him: that's what they always did with these people. He then left Hampshire alone with the prisoner. The man said immediately that he would tell Hampshire nothing unless Hampshire guaranteed that he would be turned over to the British. Hampshire replied that he could not give such a guarantee. So the man told him nothing before he was shot by the French.

Moral Feelings, Moral Reality, and Moral Progress. Thomas Nagel, Oxford University Press.
© Oxford University Press 2023. DOI: 10.1093/oso/9780197690888.003.0001

Another philosopher, when I told him this story, commented drily that what it showed was that Hampshire was a very poor choice for the assignment. But I tell it here not in order to determine whether Hampshire did the right thing in failing to lie to the prisoner in these circumstances. I offer it as a real-life example of the force of a certain type of immediate moral reaction. Even those who think that Hampshire should, for powerful instrumental reasons, have made a false promise of life to this man facing certain death can feel the force of the barrier that presented itself to Hampshire. It is an example of the sort of moral gut reaction that figures prominently in the recent literature of empirical moral psychology. I assume that a scan of Hampshire's brain at the time would have revealed heightened activity in the ventromedial prefrontal cortex.

Much intellectual effort has gone into the delineation of the protective boundaries around people that ordinary morality says we must not cross. Usually the examples designed to call forth our intuitions are more artificial than this one—as in the famous trolley problem. But the phenomenon is real and an inescapable part of human morality. I am interested in the question of how to decide what authority to give to these moral judgments, or perceptions, or intuitions—what kind of thinking can lead us either to affirm them as correct and fundamental or to detach from them so that we come to regard them as mere appearances without practical validity—or alternatively perhaps to step back from them but nevertheless allow them some influence in our lives that is not fundamental but is derived from other values. This problem has been around for a long time, and much of what I say about it will be familiar. But recent discussion prompts another look.

It is a question of moral epistemology—not the kind of epistemological question posed when we consider how to respond to a general skepticism about morality, or about value, but an epistemological question internal to moral thought. There is a venerable tradition of skepticism about whether any moral judgments, or the intuitions that support them, can be regarded as correct or incorrect, rather than as mere feelings of a special kind that we express in the language of morality. I am not going to enter into that larger debate here. I will proceed on the assumption that it makes sense to try to discover what is really right and wrong, and that moral intuitions provide prima facie evidence in this inquiry. The problem I want to discuss arises because, for some of our most powerful intuitions, there are various possible explanations, both moral and causal, that would, if correct, undermine their claim to fundamental authority—the claim that those convictions should be taken at face value as perceptions of the moral truth. Challenges of this kind present us with the task of finding a way to conduct ourselves that is consistent with the best understanding of ourselves from outside—as biological, psychological, social, or historical products.

The question has broad legal and political importance because in liberal constitutional regimes, many of the rights and protections of the individual against the exercise of collective power appear initially as intuitive boundaries of this type. Freedom of religion, freedom of thought and expression, freedom of association, sexual and reproductive freedom, protections of privacy, prohibitions of torture and cruel punishment are all supported and in part identified by an immediate sense of what may and may not be done to people, a constraint that precedes cost-benefit calculations.

Even though it is possible to construct more or less plausible consequentialist justifications—justifications in terms of long-term costs and benefits—for strict legal rules embodying such protections, that is not the moral aspect under which they immediately present themselves. The violation of an individual right seems wrong in itself, and not merely as the transgression of a socially valuable strict general rule. The question is whether this is an illusion—a natural illusion built into our moral psychology. Though Hampshire's uncrossable boundary arose in the context of an individual decision, it feels similar to that which bars the state from employing torture to get information, even against its enemies and for reasons of national security. And as we have seen recently, the bar against torture is not uncontested.

2.

I have a lot of sympathy with the classic intuitionist W. D. Ross when he says, "To ask us to give up at the bidding of a theory our actual apprehension of what is right and what is wrong seems like asking people to repudiate their actual experience of beauty, at the bidding of a theory which says 'only that which satisfies such and such conditions can be beautiful.'"[1] Still, some move into theory seems unavoidable in response to the deep disagreements that repeatedly emerge in relation to this subject.

---

1 *The Right and the Good* (Oxford University Press, 1930), p. 40.

John Rawls gave the name "reflective equilibrium" to the process of putting one's moral thoughts in order by testing general principles against considered judgments about particular cases, and adjusting both until they fit more or less comfortably together.[2] The process does not treat particular judgments as unrevisable givens, nor general principles as self-evident axioms, so it need not be conservative: it can lead to radical revision of some of the considered judgments with which one begins. But it must take intuitive value judgments as starting points, and in order to dismiss some of those judgments as mistaken, it must rely on others—just as we must rely on perceptual evidence when dismissing some perceptual appearances as illusions. I think there is no alternative to this method for pursuing answers to moral questions in which we can maintain some confidence, even in the face of disagreement.

While the process is structurally similar to that of testing empirical hypotheses in the natural or social sciences against observational evidence, there is a crucial difference. In the scientific case we understand our perceptual observations to be the result of causal interaction with the world we are investigating. Even though we haven't solved the mind-body problem and don't understand how the brain produces conscious experience, we have a rough conception of ourselves as organisms causally embedded in the world whose constituents and laws we are trying to discover; so how things seem from our perceptual point of view clearly provides data for such an investigation. However, in the moral case,

---

2 *A Theory of Justice* (Harvard University Press, 1971), p. 20.

we do not take our evaluative intuitions to be the result of causal interaction with the moral domain, and it is not clear what other kind of embeddedness or access to the moral truth moral judgment might involve.

As is often observed, moral judgment has this in common with logical and mathematical judgment, which are also not the result of our causal interaction with the realms of logic and mathematics. Knowledge arrived at just by thinking is mysterious. But I am going to leave aside the large and difficult question of how this is possible and concentrate on more specific issues.

Here are the familiar features of ordinary moral thought that give rise to our problem. We evaluate many different kinds of things, but important among them are states of affairs or outcomes, on the one hand, and actions or policies, on the other. To evaluate states of affairs, we use the concepts of good and bad, better and worse. To evaluate actions, we use in addition the concepts of right and wrong. The classical problem is whether there is an independent aspect of morality governing the rightness and wrongness of acts and policies—either of individuals or of institutions—or whether the only truly fundamental values are good and bad, so that standards of right and wrong must be explained instrumentally, by identifying the types of actions and policies that lead to good and bad outcomes. The latter possibility was given the name "consequentialism" by Elizabeth Anscombe,[3] and its best-known

---

3 "Modern Moral Philosophy," *Philosophy* 33 (1958), reprinted in G. E. M. Anscombe, *Ethics, Religion and Politics: Collected Philosophical Papers Volume III* (University of Minnesota Press, 1981), p. 36.

version is utilitarianism. The opposite view, that the right is at least in some respects independent of the good, doesn't have a name, but the principles that it identifies are usually called "deontological"—an ugly word, but we seem to be stuck with it. Deontological principles say that whether an act is morally permitted, prohibited, or required often depends not on the goodness or badness of its overall consequences but on intrinsic features of the act itself. In a case like Hampshire's, the calculation of probable consequences is clearly in favor of lying to the prisoner, so if it would be wrong to do so, that would have to be for some other reason.

The coexistence in ordinary moral thought and discourse of these two types of evaluation is the starting point for moral reflection that might lead us to a new equilibrium. If we take both consequences and deontology seriously as guides to conduct, they will naturally present us often with the sense of a dilemma. When a deontological prohibition—against killing the innocent, against breach of promise, against betrayal—blocks an act that would prevent a greater evil or produce a great good, we are likely still to feel the force of the reason to promote the good, which makes it tempting to violate the prohibition. I believe the sense of moral conflict in these cases arises naturally and is not just a philosophical artifact.

One possible response to such a dilemma is simply to take the consequentialist reasons as decisive and to decide that they should override the deontological intuition, which is construed as a form of moral squeamishness or a groundless taboo. Yet without more explanation, this all-consequentialist response would be arbitrary. The problem arises because we have both types of moral intuition, and if we are going to let one type override the other, it could just as easily go in the opposite direction: we could decide that the principle that one should always do what will have the best overall consequences is shown to be wrong by the obvious impermissibility of using certain means like torture to achieve good ends. The conclusion would be that to allow good and bad outcomes always to determine right and wrong is an illusion, perhaps an illusion about what is demanded by rationality.

Merely pitting the intuitions against each other results in a standoff. To decide which of them to credit and which, if any, to throw out we must do more. For each side of the dilemma, we must consider the pros and cons—both how to make the best sense of these types of judgments and how most convincingly to undermine their authority.

On the positive side, it is easiest to explain the meaning of consequentialist values, as in utilitarianism. Each of us has direct access to the goodness or badness of certain things in our own lives, such as pleasure and pain, freedom and coercion, survival and death. Once we accept the crucial judgment that such things are objectively good and bad, the imagination allows us to extend these values to similar features in the lives of others. There seems no reason to weigh a given quantity of pleasure or pain differently depending on who undergoes it. So the obvious way to compare the value of two states of affairs is to add up the good or bad in the lives of all those involved and see which has the higher net balance. That then is the one we should prefer and, if possible, bring about. Good is preferable to bad and better is preferable to worse. Whether or not they are correct, consequentialist value judgments are unmysterious.

The positive interpretation of deontological values is less transparent, but I believe we can give them a definite sense. When we think in this way about how to value others, it is their status as autonomous beings independent of us that is central, not their susceptibility to pleasure and pain, or other good or bad things that might happen to them. Deontological requirements govern our direct interaction with each other person; they determine how we may treat him rather than what we should want to happen to him. As with consequentialist values, this is an extension to everyone of a value whose importance we can recognize in our own case. The basic idea is that we must regard each person, including ourselves, as immune from subjection to others—the center of a morally protected sphere of individual autonomy that can be granted equally to everyone. Each of us respects this autonomy in others most fundamentally not as a good that we should promote but as a boundary that we must not violate. We must not violate it even to prevent more violations of the same kind by others—hence the prohibition against torture even to get information about a planned terrorist attack. This idea of inviolability seems clear even if there are significant uncertainties in the precise shape of the deontological boundaries. There is room for indeterminacy and ongoing reflection here as elsewhere. So far this is merely an interpretation of deontology as an intelligible way of valuing people—not a proof of its correctness. But in that respect it is on the same footing as the interpretation I have given of consequentialism.

According to this account, our deontological convictions that certain things may not be done to people are just the subjective experience of running up against objective boundaries of inviolability that define a systematic form of value. Hampshire evidently felt that to manipulate the prisoner by falsely promising him his life was something he just could not do, even to an enemy. It is phenomenologically more complex than the feelings of impartial sympathy for the pleasures and pains of others that would motivate an obedience to utilitarian requirements. But it is intelligible as a response to a different kind of value that humans possess.

3.

Thus far there is something to be said on the positive side for both types of intuitions, consequentialist and deontological. However, this is not the end of the story. I said that it would be necessary to examine the pros and cons for both types of intuition, and the most interesting arguments are those that claim to undermine the authority of some of our intuitive starting points—especially the deontological ones. Let me describe two.

The first is a theory due to David Hume, which seeks not to discredit deontological requirements but to show that they are not morally basic, because they can be explained in terms of other values.[4] The theory is called rule-consequentialism, or in one of its more specific versions, rule-utilitarianism. It purports to vindicate many of our deontological intuitions—about property, contract, promise, political obligation, and various individual rights—by

---

4 *A Treatise of Human Nature* (1739–40), Book 3, Part 2.

attributing them to our having internalized certain conventions with strict rules that serve the general interest.

The theory is presented by Hume in his account of what he calls the "artificial" virtues, and its point is that these moral requirements do not have a foundation independent of the general good of society. They seem to need an independent foundation only because the general good of society requires strict rules, such as the rules of property and contract, that can serve their consequentialist purpose only if they are followed even when their violation in an individual case would produce more good than harm. The utility of the rule in creating security and predictability demands that utility not be considered in deciding whether to adhere to it in each individual case. Participants in these conventions who have internalized the rules experience the wrongness of stealing and promise-breaking as simple and intrinsic; but those requirements do not wear their true meaning on their sleeves. Their conventionality, and their consequentialist foundations, are hidden, and this too is an aid to their effectiveness.

Although I believe it cannot give a fully adequate account of rights, Hume's theory is a brilliant contribution to moral philosophy. It offers to explain deontological requirements in a way that also opens the possibility of revision of those requirements if the good of society would be better served by modification of the governing conventions and rules. I believe rule-consequentialism is at least part of the truth about rights and deontological obligations. This is strongly supported by the fact that societies that have abandoned the protection of individual rights in order to leave their hands free for the more effective pursuit of the general good have some of the most dismal records with respect to its actual achievement.

Rule-consequentialism doesn't seek to eliminate deontology but tries to show that it is not morally fundamental. But a second, more negative criticism of anticonsequentialist moral intuitions has recently become prominent. It is found in the writings of certain psychologists, notably Daniel Kahneman,[5] Jonathan Haidt,[6] and Joshua Greene,[7] who have turned their attention to the psychological analysis of moral judgment and motivation, and also to their neurophysiological, evolution-ary, or sociological underpinnings. The approach has also been taken up by non-psychologists like Peter Singer[8] and Cass Sunstein.[9] Like Hume, it ascribes to deontological rules some form of social utility, but it is often disposed to find that utility more pronounced in the past, when our hunter-gatherer ancestors lived in small groups and evolutionary forces produced the then-functional dispositions that we find operating in us to this day. Those dispositions strongly inhibit direct interpersonal aggression and violation of interpersonal agreements, and therefore contribute to peaceful coexistence and cooperation. They are efficient because they don't require knowledge of long-term consequences, but respond only

---

5 *Thinking, Fast and Slow* (Farrar, Straus and Giroux, 2011).

6 *The Righteous Mind* (Pantheon, 2012).

7 *Moral Tribes* (Penguin, 2013).

8 *The Expanding Circle: Ethics and Sociobiology* (Farrar, Straus and Giroux, 1981).

9 "Is Deontology a Heuristic? On Psychology, Neuroscience, Ethics, and Law" (August 1, 2013). Available at SSRN, https://ssrn.com/abstract=2304760 or http://dx.doi.org/10.2139/ssrn.2304760

to the immediate character of the action and engage the emotions rather than reasoning.

In the end, the account is partly vindicating, like Hume's, but partly debunking. It does not set out to discover the moral foundation for our intuitions. Instead, it asks us to take a detached stance toward ourselves and our moral responses—to try to understand ourselves from outside, so to speak. We then discover, it is claimed, that some of our most confident moral judgments are the product not of reason but of emotionally charged instinct, created by natural selection. Only after explaining those responses scientifically can we decide which of them we should continue to allow to influence us and which we should instead dismiss as emotional illusions.

The conclusion of these psychological critics is that while the evolutionary legacy of deontological morality retains some usefulness as a set of heuristics for identifying the right thing to do much of the time—or, to put it in Kahneman's terms, as an example of "thinking fast" as opposed to the less efficient but more accurate "thinking slow"—nevertheless, the only moral standards that are valid in their own right are the rational standards of consequentialism, more specifically, some form of utilitarianism. That is the standard that underpins what is useful in the deontological aspects of intuitive morality, and it is also the standard that justifies overriding deontological intuitions when a slower calculation of costs and benefits reveals that they would lead us to do more harm than good.

Even with this alleged scientific support, however, I believe it is too early to declare victory for consequentialism.

4.

The detached biological stance toward ourselves is now a cultural commonplace. People are accustomed to thinking of their psychological dispositions as the product of natural selection, and of their minds and motives as significantly not under their conscious rational control. But this merely poses the problem of distinguishing between moral appearance and moral reality and does not yet solve it. Psychologically reductive or debunking accounts of our moral intuitions are not self-validating; they are contributions to the process of reflective equilibrium, and we have to decide whether they are more plausible than the judgments they propose to replace.

The problem of appearance and reality is at the heart of philosophy, and it is found in every branch of the subject. It arises when we step outside of ourselves temporarily and consider the way things seem to us in some respect as a psychological fact—a fact about a certain type of creature in the world. The question then is whether the best explanation of this psychological fact—employing the forms of biological, neurophysiological, psychological, and historical understanding available to us—is compatible with our continuing to affirm that things are as they appear to be from that perspective. We can pose this question whether the appearance is a sensory perception, a memory, a mathematical certainty, an aesthetic judgment, or a moral conviction.

However, it is important that when we take this step outside of ourselves, the inside point of view that we are examining does not disappear. We cannot completely withdraw from our own point of view and observe ourselves as if we were someone else. Even when we take up the external point of view, it is still ours. In the cases we are discussing, both deontological intuitions about the wrongness of murder or betrayal and consequentialist intuitions about the goodness of saving more lives rather than fewer continue to offer prima facie grounds for moral belief, and we have to decide whether what appears to the external view will justify us in disregarding some of them.

In arriving at such a judgment, we simply have to make use of moral intuitions—otherwise we could not draw any moral conclusions. So the question will be whether the external view— scientific, historical, or sociological—of our moral responses convincingly weakens the authority of some more than others. Since the responses themselves participate in this contest, the outcome is not automatic. For the kinds of individual rights we are considering, I believe the outcome of the external challenge remains just as problematic as the original moral dilemma.

When psychological and neurophysiological data and speculative evolutionary explanations are brought to bear on these judgments, reactions typically diverge, in ways that I suspect depend heavily on the reflector's antecedent predisposition for or against a purely consequentialist outlook. Some people just find utilitarianism the only rationally intelligible form of moral justification; others find it equally obvious that morality has another dimension. For the former, the fact that deontological judgments are arrived at quickly, without deliberation, and that they are associated with emotion, both phenomenologically and neurophysiologically, is a reason to discredit them as instances of moral knowledge, especially since they sometimes require us to do what is clearly irrational, namely, choose the greater evil over the lesser. The hypothesis about their evolutionary origin provides an alternative explanation of their emotional force.

But for anticonsequentialists, the immediacy and emotional force of reactions against murder, torture, betrayal, and so forth are not surprising, since they are responses to the immediate moral character of our interaction with another person rather than to broader consequences. *Of course* your ventromedial prefrontal cortex will squirm when you think about murdering or torturing someone! Anticonsequentialists also point to another psychological fact—that impartial benevolence, the motive that is supposed to ground utilitarianism, is far too weak in most human beings to support obedience to its demanding moral requirements, which makes it unsuitable as the sole basis for human morality. As for the evolutionary speculations, they hardly count as independent empirical data: they are really driven by the consequentialist moral theory and can be taken with a grain of salt.

In other words, nothing in this complex set of arguments and counterarguments forces one to abandon either the consequentialist or the deontological position. The main point, as Selim Berker has pointed out, is that disagreements over how to respond to information about the psychology and neurophysiology of moral judgment are themselves moral disagreements.[10] It is certainly legitimate to introduce these findings into the process of reflective equilibrium, but in the end, it is we who have to decide

---

10 "The Normative Insignificance of Neuroscience," *Philosophy & Public Affairs* 37 (2009).

whether they should undermine our confidence in the validity of the deontological judgments they are supposed to explain away. And in this decision, those intuitions themselves play a role. They don't automatically withdraw from the scene in response to an MRI reading. It is not a question of whether the outside view will cause the challenged intuitions to become emotionally weaker. It is a question of whether it convinces us to regard them, whatever their emotional vividness, as mere appearances rather than recognitions of moral reality. (After all, optical illusions don't go away as features of visual experience when we learn by measurement that they are illusions.) When we regard ourselves and our moral psychology from outside, we still have to decide whether the outside view requires us to withdraw our assent from what is presented as true from the inside. And the inside view is a participant in that choice.

I think it is clear from the history of this debate and the tenacity of the two sides that there are two quite different reflective equilibria to be found here, one of which preserves a significant deontological component in morality and one of which is significantly revisionary.

5.

I suggest that in light of this standoff it is unrealistic for either side to think of itself as having refuted the other. Both are viable moral outlooks. So let us ask a slightly different question: Would a move to the revisionary position count as moral progress? That is how it is usually presented, by contrast with the conservative unwillingness to abandon a visceral attachment to basic individual rights, seen as a legacy from the past.

We know that human morality changes over time, and it seems unlikely that it will ever reach a final steady state, any more than science will. Instead of saying that some form of consequentialism is the only rational moral outlook, moral revisionists might be understood as offering us a step forward in this journey of moral development. One might think of it, mutatis mutandis, on the analogy of a later scientific theory not simply refuting, but subsuming and replacing an earlier one, in a way that preserves and explains many of its results while revising others. The difference is that the moral theories are not alternative descriptions of the external world, but normative alternatives.

Some examples of moral reform can only be seen as outright refutations of the outlooks they seek to overthrow. This is true of the contemporary revolution in views about homosexuality, which has overridden a very powerful taboo in the name of individual liberty and human happiness, and seems to be on the way to dissolving the emotionally freighted sense of pollution that has for so long sustained the taboo. On the other hand, some proposals for moral progress are simply false, sometimes atrociously so. Bolshevism was justified, after all, under the color of moral progress, and the same might be said of Fascism. But there are other examples that pose the issue of progress in a different way, as an evaluative comparison between alternative conceptions of the same moral domain.

One example that can be seen in this way, and that is closely related to the issue before us, is that of property rights. The radical hope that private ownership of the means of production could be abolished in favor of common ownership has turned out to be a destructive dream. But I find utterly convincing Hume's account of property not as a basic moral right but as a convention sustained by its indispensable contribution to the collective interest of society. And so long as it provides security of possession, succession, and exchange, permitting capital accumulation, economic planning, and cooperation over the long term, the system of property rules and rights can also serve other ends, such as distributive justice. This instrumental, anti-Lockean conception of property rights has some currency, but it is resisted by a powerful libertarian strain in Western morality, which continues to have large political influence, especially in the United States. So I think the conventionalist, largely consequentialist conception of property rights, making them a vehicle for social justice, is best presented as a call for moral progress—a call to subsume the prevailing conception in an expanded one. Reform in the morality of property would mean that property rights came to be widely seen as based not on individual liberty but on the collective good. Even though individual liberty is an important value and should be protected in the exercise of those property rights that we hold under the collectively valuable property conventions, it should not be seen as the foundation of those rights and should not determine their content.

This is an example of how a moral outlook that has considerable intuitive support might be replaced by another that is deemed superior. In my view, even though a conception of property rights based largely on liberty and self-ownership embodies significant values, its replacement by a conventionalist conception that includes some protection of liberty would be a clear example of progress.

6.

But would it be moral progress if we came to see *all* deontological boundaries not as fundamental to morality, but at best as rough guides worthy of respect only to the extent that their relatively strict protection actually serves the welfare of society or humanity as a whole? I think that depends on whether it would be progress to simplify the moral point of view.

In some sense, the moral point of view requires putting oneself in everyone else's shoes and taking the separate point of view of each individual into account in deciding what to do. The question is: How?

The appeal of the consequentialist way of valuing people impartially cannot be denied. It may seem that there is no way to take the points of view of all individuals into account without merging them into a single ocean of benefits and harms, which is then valued as a whole. Yet there is another way. In deontological morality each individual faces us separately, and the inviolability of the individual facing us dominates the rival claims of those we could help by sacrificing him to the general welfare. In a sense the inviolability of that individual stands for the inviolability of all. I think we feel this even when we put ourselves simultaneously in the place of several potential victims of these two different kinds in a typical dilemma.

This individualized moral respect is something that morality can guarantee to everyone equally: the same entitlement to be treated in certain ways, the same status, the same limits or boundaries. It determines the character of our relations with one another. Each person with whom I interact presents me with the same stubborn and impenetrable moral surface that I present to him: there are certain things that may not be done by either of us to the other. And this "moral minimum" is an expression, in the structure of morality, of the demand that moral consideration should respect each individual separately. That is not the case if morality in principle permits the complete subordination of one person's interests to the greater interests of others. As Frances Kamm has emphasized, such an inviolable status is denied to *everyone* by any consequentialist system, whatever outcomes it values.[11]

The question I am asking is whether, looking at ourselves from outside, we should come to view our attachment to rights and deontology as an unnecessarily cluttered moral outlook, which grossly magnifies the claims of the person facing us and limits our rationality. Would it be progress if we no longer took the individual-centered deontological outlook and the intuitions stemming from it as fundamental moral guides?

I believe something would be lost. No doubt human morality will evolve, and perhaps it will move in a more consequentialist direction. But the completely different way of valuing individuals, by treating each of them decently come what may, and demanding such treatment for ourselves, is a vital part of our lives. Most important, it is a distinctive way of thinking about how to relate to one another, the source of our constantly developing interpretations of

---

11 *Morality, Mortality*, vol. 2 (Oxford University Press, 1995).

people's equality of status as the bearers of human rights. Without it the advantages of membership in the moral community would be seriously diminished—not quantitatively but qualitatively. Those who disagree will see me as simply begging the question— but perhaps that is unavoidable.

2 | MORAL REALITY AND MORAL PROGRESS

1.

What is the right way to think about moral progress? The history of humanity includes progress (often interrupted or reversed) of multiple kinds: scientific, technological, artistic, legal, social, economic. I am interested in the kind of moral progress that can be regarded as an advance in understanding or knowledge, not just in behavior. An advance of this kind implies some form of realism about moral truth—implies that moral propositions can be true or false independent of what we believe, so that a change in moral belief can be described as objectively correct or incorrect.[1] One kind of progress would be to give up a false moral belief in favor of a true one. Another would be to arrive at a true belief about something about which one did not have an opinion before, perhaps because the question had not arisen. Such

---

1 Anti-realists can also give a sense to the idea of moral progress, but I shall leave that possibility aside. See Catherine Wilson, "Moral Progress Without Moral Realism," *Philosophical Papers* 39, no. 1 (2010), and Philip Kitcher, *Moral Progress* (Oxford University Press, 2021).

Moral Feelings, Moral Reality, and Moral Progress. Thomas Nagel, Oxford University Press.
© Oxford University Press 2023. DOI: 10.1093/oso/9780197690888.003.0002

changes can occur either in the attitudes of an individual or in the shared attitudes of a community. But what does it mean to describe such a moral change as progress—as an advance in our grasp of the truth?

The question can be sharpened by comparing moral progress with scientific progress. On a realist understanding of science, scientific knowledge reveals the truth about a natural world that exists independently of us and our beliefs about it. A scientific discovery in physics, chemistry, or biology reveals something about the world that was true long before we discovered it, and would have been true even if we never discovered it. If humans had never developed chemistry—indeed, even if humans had never existed—it would still have been true that salt is sodium chloride. Scientific progress, at least in the basic natural sciences, consists in the discovery of truths that have been true all along, because the world has existed with these same basic characteristics all along. Moral truth, however, if there is such a thing, is not about the natural order—about the structure of the world, or its composition, or what happens in it.

What then is it about? If it were about a Platonic realm in which moral truths have their being, metaphysically separate from the natural world and the people in it, but to which we somehow have cognitive access, then it might follow that moral truth is timeless, and that moral progress, like scientific progress, is the discovery of what has been true all along. But realism about morality, as I understand it, does not imply such a metaphysical picture. Instead, we should think of morality as an aspect of practical reason: it concerns what we have certain kinds of reasons to do and not to do. We should not think of those reasons as like chemical elements waiting to be discovered. Rather, facts about reasons are irreducibly normative truths about ourselves and other persons, and realism is simply the position that their truth does not depend on our believing them. Moral knowledge, on a normative realist view, is a species of belief about what we have reason to do that is in accordance with such truths.

Sometimes we will want to say that moral progress consists in the coming to acceptance of a moral belief that was true all along, or the abandonment of a belief that was false all along; but since these are truths about practical reason and not about the natural order, our understanding of how this is so must be different from what it is in the scientific case. If the normative domain is a domain of reasons, moral progress must be identified with a change in moral outlook that there is reason to adopt. If that reason has existed for a long time before being recognized, then we can say that we have arrived at a new moral belief that was true all along. But this will not be true in every case. Any reason for a change in moral outlook is tied to the persons and the time in moral history to which it is assigned. The existence of a reason in the present need not imply that the same reason has existed at all times. Whether it existed in the past is a further question, whose answer will depend on the circumstances then, and what considerations were available or accessible to persons at that time.

The contrast with science is stark. In 400 B.C. no one could have even understood, let alone had reason to believe, that salt was sodium chloride. It would require a long path of scientific theory and experiment to reach the conceptions of elements and chemical compounds that allowed such a proposition to be formulated and confirmed. Yet it was as true then as it is now. Its truth is in no way dependent on there being any reason to believe it.

By contrast, the truth of a moral proposition cannot be distinguished from there being a reason for people to conduct themselves in the way it prescribes. Furthermore—and this is crucial—I agree with Joseph Raz that there cannot be such a reason unless it is *accessible* to those to whom it applies. "Morality is inherently intelligible," says Raz, and "the intelligibility of morality requires that morality should be in principle accessible to the people to whom it applies."[2] In other words, a moral reason can apply to them only if they have reason to believe that that reason exists. To say this is not to say that they must actually recognize it for it to exist: that would be incompatible with realism. Sometimes people have reasons not to believe in the existence of moral reasons that actually exist, or reasons to believe in the existence of moral reasons that do not actually exist—reasons of religious authority, for example. But when such a situation obtains, the morally correct position must be supported by reasons that are accessible—reasons that those same people have some reason to accept, even if they do not acknowledge them at the moment. Reasons cannot be completely hidden, as was the chemical composition of salt in the distant past. A path to their recognition must be available *at that time* for the persons to whom they are assigned—not just through centuries of historical development.

---

2 Joseph Raz, "Moral Change and Social Relativism," *Social Philosophy and Policy* 11, no. 1 (1994), 156. See also Raz, "Can Moral Principles Change?" in *The Roots of Normativity* (Oxford University Press, 2022). The position I defend here has much in common with his point of view in those two essays.

2.

What interests me with respect to moral progress is whether we can distinguish those advances in normative understanding that constitute discoveries of what was true all along, because the relevant reasons were long accessible, from other advances that cannot be so described, because the relevant reasons became accessible only recently.

Because my interest is specifically normative progress, I want to set aside one way a moral truth may become accessible that was not accessible before: simply by a change in the non-normative facts or in our knowledge of such facts. This may pose new moral questions, questions that did not have to be addressed before. For example, it becomes clear that the burning of fossil fuels is contributing to a potentially dangerous increase in global temperatures; or advances in medicine pose questions about end-of-life decisions or parental surrogacy; or the development of a modern market economy poses questions about the legitimacy of agreements that block competition. However, this kind of factual or informational change does not by itself imply the kind of change in accessibility of moral reasons that I am talking about, since it might involve no new norms but only the application of already accessible norms to new facts.

That would be the general form of moral progress if, for example, there were a single governing moral principle, such as utilitarianism, recognizable by any rational being, from which all moral reasons were derived. Then changes in the known consequences of what people were able to do would generate corresponding changes in what they ought to do. But the answer would not depend on discovering new moral reasons of a kind that hadn't been accessible before. It would depend only on what principles of conduct, or policies, or institutions were justified by the single standard of impartial welfare maximization.

If moral reasons were like that, moral progress would in a sense resemble scientific progress, because the right answer to a newly posed question, about how to respond to climate change or medical advances, would have been true hypothetically long in advance. That is, it would have been true before anyone thought of it that *if* certain forms of artificial life support were to become available, then it would/would not be permissible to turn off the respirator/remove the feeding tube from a patient in a vegetative state under such and such conditions of prior consent. In other words, all moral progress would be the discovery of moral truths that in a sense had been true all along, like the chemical composition of salt—because they followed from a premise that any rational being could have recognized. Their normative basis—the underlying reason—would always have been accessible. But since I do not believe that all moral reasons derive from a single utilitarian axiom, I shall leave this possibility aside for now.

My concern is how to interpret moral progress if we assume instead a more pluralistic and complex conception of moral reasons, though nonetheless a realist one. I believe there will be cases where reasons come to be recognized—either in response to the presentation of new choices by new non-normative facts, or as a result of moral reflection that revises or extends existing moral attitudes—that could not have been recognized much earlier, and where the discovery is specifically normative and not merely the application of timeless norms to new nonnormative facts, as with utilitarianism.

To be a realist about such reasons is to hold that judgments made in such circumstances can be correct or incorrect, and that their correctness or incorrectness consists in their appropriateness as responses to those very circumstances. The most basic truth here, in other words, may be local—that this is how we should now move forward. Whether it was always wrong to think anything else is a different question, the answer to which depends on what there was reason to think in other circumstances, in the past. Sometimes moral progress will be presentable as the discovery of what was true all along; sometimes it won't. In the latter case, it will be because the recognizability of such truth (like the progress of scientific knowledge) is path-dependent: that a certain policy or practice would be an improvement may be understandable on reflection only by those who have already passed through certain prior stages of moral thought and practice.

We have seen many historical examples of moral progress, usually connected with political and institutional progress: the abolition of slavery, the replacement of aristocracy by popular sovereignty, the growth of religious toleration and freedom of expression, the elimination of cruel punishments, the emancipation of women, the abandonment of racial discrimination, the attempt to create equality of socio-economic opportunity, the defense of sexual freedom, improvement in the treatment of animals. Change occurs both through the work of conscious moral reformers and moral or political theorists and through disseminated alteration in the attitudes of large populations, at various velocities. To call these changes progress is to make a normative claim: that there were moral reasons to replace the practices prevailing at the time with something else. In some cases, those reasons were accessible long before they came to be widely recognized, but not always. For example, I believe the moves toward constitutionally limited democratic government and state-supported equality of social and economic opportunity required both moral antecedents and certain levels of political, economic, and educational development before they were even imaginable as options. I believe that in the absence of those conditions the reasons for those reforms were not accessible in the pre-modern world; and they did not apply to people to whom they were not accessible.[3]

3.

I shall discuss some examples in detail later on, but first let me say more about the conception of moral truth on which all this depends. As I have said, it is a form of normative realism that is essentially local. Reasons do not appear in a description of the natural world, but the rational beings who have those reasons do, so reasons do not exist in a separate, Platonic realm. Reasons are reasons for *people* to do things, and do not exist apart from rational

---

3 For completeness, let me say that I would apply the same analysis to the phenomenon of moral decline. However, in all actual cases of moral decline or regression, such as the Bolshevik and Nazi revolutions, the moral truths that are no longer accepted remain accessible and do not cease to apply to those who reject them. It would require a catastrophic loss of cultural memory and descent into barbarism to return a society to a condition in which moral truths once recognized were no longer accessible.

beings. Although moral principles are often stated in a way that leaves out these subjects, they are always there by implication. "It is wrong to give a false promise, knowing you do not intend to keep it" is a statement about what some implied class of individuals have a special reason to avoid.

This implies that if one goes back far enough, one will reach a time when every moral principle, or principle of practical reason, however elementary, ceases to be true—simply because there is no one for it to be true of. The fact that an action will cause pain is a reason not to do it; but in this view, that was not true before the appearance of life in the universe. But I would go further, and say that it was not true until there were creatures capable of grasping the general concept of a mental state like pain which could be experienced either by themselves or by other beings—the concept of other minds, in short. That is a necessary condition of the reason being accessible to them, and therefore of its applying to them.

We can observe something similar without going into the distant past, if we consider what it makes sense to say in the present about creatures that have not reached the level of rationality needed to have access to certain reasons. Migratory birds fly south for the winter, but they do not do so for prudential reasons—that they will freeze or starve if they stay put. They do not have such reasons. We can say of a human being that he has reason to do something now because if he does not, he will suffer or die in the future—and we can say this even if he does not know the predictive conditional, since he could come to know it and could appreciate the reason that it gives him. He has the necessary conception of his future self and the capacity for prudential rationality that derives from it. But a creature that lacks this conception and capacity does not have such reasons, in objectively similar circumstances. I believe it would be a mistake to say that there are reasons for such a creature to act prudentially, but that it is incapable of appreciating them. The creature's capacity, in principle, of appreciating them is a necessary condition of their existence. That is simply a consequence of the nature of normative truth.

The same could be said of creatures like ourselves in the course of our gradual development into rational beings. At birth we do not have reasons of any kind, and we develop only over time the conceptions of ourselves, of the future, of the consequences of our actions, and of the reality of other persons that are needed to make even the most basic prudential and moral reasons accessible to us. The details of the process by which children become adults to whom the full content of morality applies are complicated and of great interest and importance, and I will have more to say about the topic later.

Finally, as in other domains where ontogeny recapitulates phylogeny, over its history our species must have evolved from the moral equivalent of early childhood through the development of conceptions and forms of thought that gradually made the reasons of prudence and morality accessible, and therefore made it the case that we had those reasons.

One might ask whether this can really be described as a form of realism about the normative domain. In holding that accessibility or recognizability by an individual is a necessary condition of a reason applying to that individual, am I not making reasons mind-dependent or response-dependent in a way that is incompatible with realism? My answer is that the way the existence of reasons depends on the mental nature of the individual that has them is not the same as the response-dependence that character-izes anti-realism. This is a fine line, but I think it can be drawn.

If my having a reason depended on my believing, or being dis-posed to believe, that I had it, or being disposed to act in accord-ance with it, that would be anti-realism. It would amount to reducing a normative truth to some kind of psychological truth rather than giving it an independent and irreducible status. But that is not what I have proposed. The accessibility that is a nec-essary condition of a reason's applying to an individual consists in the subject's being capable of understanding that he has that reason—not his actually believing it. In every case, the content of a reason is determined not by the person's attitudes but by the facts that generate that reason for someone capable of under-standing it—facts about the effects of his actions on others, for example. It is because such understanding requires certain con-ceptions and reasoning capacities that the reason does not apply to him unless he meets certain mental conditions: it is an instance of the principle that "ought" implies "can." This is a kind of mind-dependence that I believe is compatible with realism, since—provided he meets the mental conditions—the reason applies to him whether or not he accepts it.

My realism about the normative is simply the view that nor-mative propositions, to the effect that something is a reason for action or belief, are among the types of things that can be true just in themselves, without having to be analyzed in terms of any other type of truth. Nothing else *makes* it true that there is a reason to perform an action that will alleviate pain; it's just true. Not every-thing can be analyzed in terms of something else: whatever one's philosophical worldview, there have to be some types of truth that are just true in themselves—mental facts, physical facts, mathematical facts, social facts, whatever. Most of the great disputes in philosophy are over where, in this search for ultimate grounds, the buck stops. I believe normative reasons are one of those irreducible stopping points. While the existence of a reason will often require explanation, the explanation will always depend on other reasons. The question "What *are* reasons?" seems to require an answer only if you believe that the stopping points are limited to some other kinds of truth, such as physical or psychological or historical.[4]

There is one further problem about the relation to time of the local, person-dependent realism I have proposed. I have said it implies that moral truth is not timeless, because it requires the existence of persons to whom the operative reasons apply and to whom they are accessible. But one might object that if it is true now that it is wrong for rational beings like ourselves to make false promises, it was already true, even before there were any rational beings, that in the future there *would be* rational beings to whom that moral principle would apply. Moreover, even if the universe had never contained any rational beings, would it not still have been true that *if* there were rational beings of the appropriate kind, that principle and the relevant reasons would apply to them? If so, wouldn't that mean that moral truth is timeless after all, so that in a sense, moral progress is always the discovery of what was true all along?

This is a tricky question. Though I am not confident, I am inclined to answer it in the spirit of Aristotle's *De Interpretatione*.

---

4 The best statement and defense of this position is T. M. Scanlon, *Being Realistic About Reasons* (Oxford University Press, 2014).

If a sea fight occurs on Tuesday, it follows that the statement made the previous Sunday, "There will be a sea fight on Tuesday," was true. But what makes it true is what happens on Tuesday and not anything that was the case on Sunday. The truth of all future tense statements predicting the sea fight is simply a logical consequence of the event on Tuesday. I think the same is true of the prediction that there will be persons of a kind to which certain moral reasons apply. Moreover, and more important, I think something similar is true of the conditional truth that if there were persons of that kind, they would have those reasons. There is nothing in a universe that never has and never will include rational beings that makes that conditional true. The conditional, if true, would be true in that possible world only because the reasons apply in light of certain actual conditions in this world.[5]

An analogy: In a possible world in which life never appears, is it nevertheless true that if there were cats, they would be mammals? I am not sure. But if so, there is nothing about that possible world that would make the conditional true. If it is true, its truth is simply a logical consequence of the fact that cats are mammals in the actual world.

Finally, a word about the centrality of reasons in this discussion. I have framed the issue as one about reasons for action because my topic is moral progress, and I believe that the answers to moral questions of right and wrong depend on such reasons. However, it would be compatible with my normative realism to

---

5 Cf. David Wiggins (interpreting Hume): "The human scale of values is timeless or (if you prefer) it reaches backwards and forwards *to all times*." *Ethics: Twelve Lectures on the Philosophy of Morality* (Harvard University Press, 2006), p. 374.

acknowledge that at least in some cases reasons may not be fundamental but may depend on more basic values of a different kind, to
which the accessibility condition does not apply. Good and Bad,
as evaluations of states of affairs, are obvious examples. Perhaps
"suffering is bad" and "pleasure is good" are necessary truths,[6] not
analyzable in terms of reasons, and it is a normative consequence
of these truths that any rational being has reason to want suffering (not just his own) to stop and pleasure to continue. To insist
that reasons are fundamental even in these cases—that good is
just what there is reason to promote and bad is just what there
is reason to prevent—would have the unwelcome consequence, if
I am right about the accessibility condition for reasons, that nothing was good or bad before there were rational beings, so that the
suffering of animals wasn't bad in the pre-human past.

One could avoid this result by saying that when we describe
that distant suffering as bad we are speaking from the point of
view of the reasons we would have to prevent it, given our rational
nature. But I would prefer to be more pluralistic about the domain
of value and admit that not everything in it can be reduced to reasons. Perhaps some things just are good or bad in themselves, and
this is the source rather than the consequence of some of our reasons for action. There will still be other reasons that are fundamental, including many that are important for morality. In particular,

---

6 Sharon Hewitt Rawlette, in *The Feeling of Value* (Dudley & White, 2016) makes the
very interesting claim that Good and Bad are simply basic phenomenological qualities
of certain conscious experiences. Pleasure feels good and pain feels bad. She holds that
these properties are essentially and inseparably both phenomenological and normative,
and that they are the foundation of all normativity.

I remain convinced that morality cannot be explained entirely, or even largely, by reasons deriving from the goodness or badness of states of affairs, in the consequentialist vein of utilitarianism.

But this has been a long logical-metaphysical digression. The morally interesting questions about the temporality or locality of moral truth, and the correct understanding of moral progress, concern what is true at different stages in the historical development of real human beings and their institutions.

4.

I have proposed that if we go back far enough in time, none of the moral or other normative truths that we now believe were true (except in a logically derivative sense), because there were no rational beings around to have the reasons in question. But if we put that aside and instead consider the human past, it becomes an open question with regard to any new moral belief whether, if true, it is the discovery of something that has been true for a long time—or whether it is the discovery of something that became true more recently, because the relevant reasons were not accessible much earlier. Let me illustrate the distinction with a pair of examples.

The clearest cases in which moral progress is the recognition of something that was true all along are cases where a great injustice is overthrown. The reasons for the change are not subtle, and typically they have long been obvious to some people (and not only to the victims of the injustice), even if most have been oblivious to them. An example that we have lived through is the recent dramatic and rapid change in the status of homosexuality in most Western countries. It began with a campaign to get the state out of the business of enforcing standards of personal morality—as opposed to preventing conduct that harms others—by arguing that voluntary sexual conduct is a private matter. This argument deliberately abstained from requiring those who thought homosexuality immoral to change their opinion: it asked them only not to use law to enforce that opinion. The strategy had some success in decriminalizing homosexual conduct, but the appeal to mere toleration was not satisfactory, and it came under assault with the movement for gay rights and the massive refusal of homosexuals to hide any longer. In spite of opposition by some religious communities, the view that there is nothing wrong with homosexuality made extraordinarily rapid progress, driven by the discovery, once the closet was flung open, that almost everyone had friends, relatives, and colleagues who were gay and whom they could not regard as moral outcasts. Heterosexuals who had been formed in a homophobic culture were able, amazingly, to make a rapid switch from being unable to imagine the erotic lives of homosexuals without fear and disgust to recognizing this as another form of human love. And the young coming of age in the new climate often couldn't even understand the old taboos.

This is a case of moral progress that thoroughly discredited the old outlook. The biological function of sex as the means of procreation provided no reason to condemn or be ashamed of sexual desires that could not result in procreation, and the reasons of individual freedom and happiness against blocking the fulfillment of those desires were overwhelming. The suppression of homosexuality had been wrong all along, and the advance in moral opinion simply recognized this. The crucial point is that the reasons against it were in principle widely accessible for a long time, even though most people were blocked from recognizing them by ignorance and social conditioning. This is shown by the rapidity with which those reasons were accepted by millions of people once the blocks were removed, without the need for gradual change over generations.

Of course, access to those reasons depends on the capacity to recognize the freedom and happiness of individuals, whoever they are, as relevant values in the justification of moral requirements. That capacity has not always existed, but I assume it is quite ancient, and that it substantially antedates the development of modern secular morality. It was certainly available long before the recent revolution in attitudes toward homosexuality.

It may be that the next stage of reform, the legalization of same-sex marriage, is an example of the second type of moral progress—an advance at a particular historical time whose identification as an advance depends on reasons that did not exist, and that could not have been recognized, very much before the issue was raised in the context of the gay emancipation that preceded it. But I won't take up that question and will instead illustrate the second type of progress with a different example, that of freedom of expression.

While there is disagreement in contemporary democratic societies about the exact boundaries of what should be protected, there is broad agreement that the state's authority to restrict what may be said or published is very limited. The presumptive right to freedom of expression can be overridden only in exceptional cases and on narrowly restricted types of grounds—never merely because the benefits of doing so are thought to outweigh the harms. This specific moral conviction about the limits on what the government may do with its power over the individual is the consequence of a broader conception of the relation between the state and its citizens, namely, that a government is legitimate only if its authority can be recognized by citizens who at the same time regard themselves as autonomous rational agents.

This is a conception of popular sovereignty whose institutional embodiment is accountability of the holders of state power to the consent of citizens through democratic elections. That in turn requires that their consent not be manipulated by the exercise of that power to control what they may hear and read. So there is a direct connection between a modern conception of political legitimacy and recognition of a right to freedom of expression. But it doesn't stop with political speech: the conception of citizens as autonomous rational agents leads to a broad interpretation of the range of protected expression and also to a more general resistance to encroachments on individual autonomy by the state.

This means, I think, that recognition of a strong right to freedom of expression has become accessible only in the modern era and is therefore an example of the second type of moral progress—not the discovery of a moral truth that had been true all along. It was not accessible, and therefore not applicable, in pre-modern times because the reasons behind it are intelligible only to those who understand from the inside the conception of political legitimacy on which they depend. Someone who believed that the authority of rulers depended on a divine source, or on dynastic right, or who believed with Hobbes that those who command a monopoly of force deserve our allegiance as the only protection against the horrors of anarchy, would not be able to understand this kind of restriction on the grounds for the exercise of their power. Such a pre-liberal subject might still find fault with legal restrictions of expression on the ground that they have bad consequences or are based on false beliefs, but he would not be in a position to criticize them for relying on justifications that undermine the acceptability of the state to autonomous subjects.

I am claiming, in other words, that one needs already to have arrived at a modern understanding of the conditions of political legitimacy and the autonomy of the individual in relation to the state, in order to be able to engage in the reasoning that allows one to see what it entails with regard to freedom of expression. This is moral progress, and it should be understood realistically—that is, as the discovery of objective reasons to adopt this new principle. But the reasons did not always exist, because the conditions for understanding them were not always present.

There is a natural objection to this claim: Couldn't someone with highly developed rational faculties living in the England of Henry the Eighth—for example, Sir Thomas More—have come to see that the regime lacked legitimate political authority because that authority could not be recognized by citizens who regard themselves as equal, autonomous, rational agents? And couldn't he then have gone further to draw the conclusion that a government that was legitimate in that sense would not have the authority to restrict freedom of expression on consequentialist grounds (so that his own persecution of heretics was impermissible)?

I do not think so. Such thoughts do not depend only on the exercise of a universal faculty of reason but require also that the thinker have a conception of the relation of the individual to the state that has been shaped by actual historical developments and the experience of living under institutions that claim secular authority of the kind in question. Thomas More did of course challenge the authority of the king, but only on the basis of a higher authority—not in light of his status as one of the autonomous rational agents to whom the government is accountable. I also think that More could have come to recognize that his persecution of heretics was wrong—by recognizing that his grounds for certainty in his religious beliefs were insufficient to warrant sending people to the stake. But the liberal conception of freedom of expression was out of his reach—and of course out of the reach of his less sophisticated contemporaries.

In a way I am conceding a criticism that Bernard Williams made against me regarding the universalist pretensions of political liberalism—though I do not accept the "relativism of distance" that he favored as an alternative. He memorably quipped:

> Must I think of myself as visiting in judgment all the reaches of history? Of course, one can imagine oneself as Kant at the Court of King Arthur, disapproving of its injustices, but exactly what grip does this get on one's ethical thought?
>
> In particular, is it really plausible that one makes this imaginary journey only with the minimal baggage of reason? Granted the notable fact that no one had the liberal worldview then, the ethical time traveler must take with him implicitly the historical experience which has made him the

liberal he is, and that experience does not belong to the place he is visiting.[7]

In acknowledgment of this point, I am offering a conception of the relation between moral realism and moral progress that requires objective existence of the reasons for a positive change in moral outlook only at the time it occurs. Sometimes progress will occur as a result of the recognition of reasons that have existed unrecognized for a long time; but not always. The claim of moral progress is essentially a moral comparison between two available alternatives, not a comparison of both of them with some independent moral reality. The question is, "Where should we go from here?" and one can be as much of a realist about emerging moral reasons as about timeless ones.

5.

At present we face difficult moral problems in the domains of socioeconomic inequality, global justice, biomedical policy, and our obligations to future generations. I believe a philosophical understanding of moral progress will help us to think about the future and will allow us to take an appropriately critical attitude toward beliefs that may seem settled in the present. But in order to explore further the accessibility condition for moral truth

---

7 Bernard Williams, Review of *The Last Word* by Thomas Nagel, *New York Review of Books* (1998); reprinted in Williams, *Essays and Reviews 1959–2002* (Princeton University Press, 2014), p. 384.

that I have proposed, it will be useful to say a bit more about the past.

Sometimes, as with animals and children, failure to meet the accessibility condition is due to their biologically given nature, or their nature at that stage of their lives. And it is possible that the biological nature of human beings will evolve in the future to give them access to new forms of practical reasoning and morality. But I am interested here in differences in accessibility that are due to historical and conceptual changes. To recur to the example of freedom of expression, it is obvious that Thomas More was not biologically blocked from recognizing the liberal conception of political legitimacy. What he lacked was the historical experience of constitutional democracy and popular sovereignty and the conception of the autonomous individual that informs them. As Williams says, the judgments that can be made by persons who have lived under such institutions are not the manifestation of a completely general practical reason that could be applied with the same result from any historical point of view.

So there is a clear sense in which the moral principles associated with liberal constitutional democracy were not accessible to More. In another, weaker sense, one might say they were accessible: his cognitive capacities were presumably such that if he were transported to the present and had those principles explained to him and been able to observe them in operation, he could have come to understand them. But on the view I am proposing, this is not sufficient access to permit the principles to apply to him.[8]

---

8 I should mention a natural alternative to my view, namely that the stronger condition of accessibility that I favor is not a condition of the application of moral principles

Let me turn to a more radical example, prompted again by Bernard Williams—that of ancient slavery. The current condemnation of slavery as a paradigm of injustice is usually thought of as a timeless moral truth: slavery has been unjust for as long as it has existed, and it is hard to understand how anyone could ever have thought otherwise. With regard to Negro slavery in the Americas of the seventeenth, eighteenth, and nineteenth centuries, it seems clear that the reasons for its injustice were accessible from the start, and that the gradual growth of abolitionist sentiment can be regarded as a widening recognition of what had been the moral truth all along. Surprisingly, however, this is less obvious in the case of ancient slavery. In chapter 5 of *Shame and Necessity*, Williams discusses attitudes to slavery in ancient Greece and asks "how far our rejection of that institution . . . depends on modern conceptions that were not available in the ancient world."[9]

Williams writes:

The Greeks had the institution of chattel slavery, and their way of life, as it actually functioned, presupposed it. (It is a

---

to an individual, but only a condition of blaming him for violating them. On this view, the liberal principle of free expression applied to More (since he was not cognitively incapable of understanding it), so that he was wrong not to respect it. But because the principle was not historically accessible to him, he is not blameworthy for this failure. While I acknowledge that this is a possible position, I reject it because in my view the judgment of wrongness depends on reasons for action which, in the world as it was, More did not have. I recognize that the choice between these two alternatives is not obvious.

9 Bernard Williams, *Shame and Necessity* (University of California Press, 1993), p. 106.

different question whether as an abstract economic necessity they needed it: the point is simply that, granted the actual state of affairs, no way of life was accessible to them that preserved what was worthwhile to them and did without slavery.)[10]

Slavery, in most people's eyes, was not just, but necessary. Because it was necessary, it was not, as an institution, seen as unjust either; to say that it was unjust would imply that ideally, at least, it should cease to exist, and few, if any, could see how that might be.[11]

It is only Aristotle who took up the question, and he tried to show that slavery was just, prompting Williams to remark that "if there is something worse than accepting slavery, it consists in defending it."[12]

The other element besides necessity in the prevalent Greek attitude to slavery was luck. It was acknowledged that to be a slave was a piece of disastrously bad luck, and its arbitrariness was evident because misfortunes of war or piracy could in principle result in anyone's becoming a slave, if they fell into the wrong hands. The conception of slavery as a necessary institution, and bad luck as the determinant of who suffers the misfortune of being a slave under that institution, combined, in Williams's interpretation, to cut most Greeks off from being able to think about its justice or injustice. Those two concepts took the question of its justice off

10 Ibid., p. 112.
11 Ibid., p. 117.
12 Ibid., p. 111.

the table, so to speak. This, Williams says, is precisely where the gulf lies between our moral conceptions and those of the Greeks:

> Obviously we do not apply those concepts [necessity and luck], as the Greeks did, in such a way that we accept slavery. But we do apply those concepts very extensively to our social experience, and they are still hard at work in the modern world. The real difference in these respects between modern liberal ideas and the outlook of most Greeks lies rather in this, that liberalism demands—more realistically speaking, it hopes—that those concepts, necessity and luck, should not *take the place of* considerations of justice. If an individual's place in society is to be determined by forces of economic and cultural necessity and by that individual's luck, and if, in particular, those elements are going to determine the extent to which he or she is to be (effectively, if not by overt coercion) in the power of others, then the hope is that all this should happen within the framework of institutions that guarantee the justice of these processes and their outcome.[13]

This is very convincing, but does it show that the conceptions that lead us to condemn slavery as unjust "were not available in the ancient world"—so that what is true now was not always true?

It does seem that the conception of justice as an all-embracing value, which applies to the whole social, economic, and legal structure of a society in light of its impact on people's lives, is a

---

13 Ibid., p. 128.

relatively recent invention. But one would think that a more narrowly focused sense of the unfairness of the individual relation between master and slave, with its huge and undeserved inequality of power and well-being, would have been available nevertheless to anyone who was able to imagine himself in the place of another. Or is this just a projection into the past of a universalist idea—the claims of humanity as such—that would have been unavailable to the Greeks as the ground for indicting a basic social institution?

I do not find it credible that such an idea was inaccessible to them. And if the Greeks were able to recognize that the interests of human beings as such had any value at all, they could also have recognized that the crushing power of masters over slaves in their society was so great that there must be something wrong with it— even if they could not imagine a viable social structure without it. This type of inchoate moral uneasiness is an important factor in the development of moral and political thought, and continues to be so today. Williams may be right that the modern conception of justice as a comprehensive virtue of social institutions, according to which slavery is self-evidently unjust, was not available or accessible to the Greeks. But it is hard to believe that they did not have access to the sense that there was something sickening about slavery, something that demanded an effort of the moral imagination to conceive of alternative forms of life rather than to take it as a given.

The most accessible form of such an insight, expressed as a moral principle, would have been the recognition of a universal claim to the most basic kind of individual autonomy. This would mean that there are certain forms of absolute domination and denial of control over one's life to which no one should be subjected. It does not imply any more general requirement of equal treatment or impartial consideration, and therefore might well have been accessible much earlier, even in the ancient world, to anyone with a robust interpersonal imagination. Perhaps Aristotle's defense of the justice of slavery on the ground that some people are naturally incapable of governing themselves is a sign that he was not immune to the appeal of such a universal norm and had to find a way to resist it.

6.

I said earlier that there are two very different ways of understanding moral progress, depending on different conceptions of the structure of morality. One is to see progress as resulting from the gradual recognition and application of a single timeless normative principle such as that of impartial benevolence. The other is to see it as the gradual development of a pluralistic system of norms and values, some of which become available only after earlier stages, both normative and institutional, have been traversed. My account is of the second kind; but whichever account one favors, it is clear that human morality has a history, and the two types of account will probably start out with a common story of its beginnings.

The minimal necessary conditions for a specifically moral form of practical reason are the capacities to (1) recognize reasons for action that apply equally to oneself and others, (2) object or feel indignation when others do not act in accordance with those reasons, (3) recognize the objections of others when one fails to act in accordance with them oneself, and (4) acknowledge such failure on one's own part with something like guilt. But these conditions can be met in many different ways, with respect both to the content of the reasons and to the scope of the class of persons to whom they apply. A norm of conduct may be recognizably moral without being remotely universal, and it need not be concerned with the general welfare.

There is now a considerable body of work, speculative but partly based on empirical observation of the social behavior of the great apes, about the evolutionary origins of human morality. The biological explanation of altruism toward near kin provides a starting point, but there is also evidence of innate dispositions not dependent on kinship that make cooperative behavior possible, with collective ostracism of free-riders or violators of useful norms. That such dispositions in humans are the result of biological and not only cultural evolution is also demonstrated by their appearance in very young children, during the first years of life.

These proto-moral dispositions, however, govern the interactions of familiar individuals with one another, either in pairs or in small groups. They do not extend to relations with outsiders, let alone to a set of universal norms. But even leaving the issue of universality aside, something more is needed for these dispositions to generate moral standards, however restricted in scope. They must evolve from mere dispositions into norms—rules or standards whose validity persists even if they are violated, and even, up to a point, if they are not recognized. What this means is that humans, as they developed new cognitive and motivational capacities, were presented with new questions of what to believe, what to want, and what to do, to which there were right and wrong answers—answers they could understand, that specified what, given their capacities, they had reason to believe, want, or do.

Probably the earliest, pre-human development was the transition from conditioned response caused by past experience to inductive reasoning based on memory, and instrumental reasoning based on awareness of prospective cause and effect. The behavioral evidence that a creature has made these transitions would come from flexibility in response to new situations that cannot be explained by mere conditioning, as when a chimpanzee uses a tool for a new purpose or imitates what another chimpanzee has done to bring about a desired result.

A big step beyond instrumental reason is the development of a conception of one's own future, together with the ability to imagine and evaluate its possibilities, which provides the basis for prudential reasoning. The information made available by these cognitive capacities presents a creature that possesses them with questions of practical reason, questions of what to do or prefer, that have right and wrong answers. And the right answer in any particular case does not depend on the conclusion that the creature actually draws but on general normative principles—as is also the case with inductive, deductive, and instrumental reason. It is this gap between truth and performance that makes the answer normative rather than merely psychological, in spite of the contingent condition of accessibility.

Normative principles apply only to beings with the capacity to understand them, but their correctness cannot be defined in terms of those capacities, because the capacities are defined in terms of the principles. We can see this from the fact that not just any cognitive or practical disposition in a creature would give rise to a corresponding normative principle. The reasoning of which the creature is capable has to be correct. The condition of accessibility does not determine the content of the norms. It merely determines their reach by specifying the kinds of beings to which the relevant reasons apply. The reasons so specified are valid in themselves and not reducible to anything else.

The cognitive and motivational capacities that are needed to understand specifically moral reasons seem to be present only in humans, and they are found in very young children, revealing an innate endowment. Children one to three years old are capable of forming joint intentions with others, appreciating the intentions, perceptions, and desires of others, and appreciating the beliefs of others about their own intentions and perceptions. They are also disposed to endorse an equal division of the product of joint action in obtaining some benefit. They are motivated to be helpful to others, and they react positively to others who are helpful to third parties and negatively to those who are harmful. The capacities to see oneself as one individual among others, to recognize social norms that apply generally to the group of which one is a member, to disapprove violations and to react with something like guilt to one's own failures to treat others correctly, follow not long after.[14]

---

14 See Michael Tomasello, *A Natural History of Human Morality* (Harvard University Press, 2016) for a fine account of research in this field, on both humans and other primates.

7.

How do we get from these humble interpersonal beginnings to the complex modern morality of political and social justice, individual rights, collective obligations, humanitarian standards, and so on, with its universal ambitions, global reach, and partly impersonal character? Although the terms of this morality are contested, I think it is not inaccurate to think of it in the singular—as a connected set of questions with a range of possible answers that are debated among members of the enormous modern moral community.

The development is partly due to the "expanding circle"[15] of concern beyond the immediate social unit, to include eventually all of humanity and even other creatures. But this would be sufficient to provide a complete account of the normative basis of moral progress only if impartial benevolence were the basis for all of morality, including those complex requirements and permissions that are not, on the surface, consequentialist in form. The utilitarian tradition has a long and honorable record of trying to subsume all of morality under a single principle along those lines, through the consequentialist justification of the different rules, practices, conventions, dispositions, and institutions in which contemporary secular morality is embodied. According to this account, progress has come through the creation over time of social, political, and economic forms of life that advance the aggregate general welfare.

---

15 Peter Singer's phrase; see Singer, *The Expanding Circle: Ethics and Sociobiology* (Farrar, Straus & Giroux, 1981).

If this were the correct account of moral progress, then the normative condition for access to any moral truth whatever would have been met when humans were first able to appreciate that the happiness or unhappiness of all human beings is of equal value, from an impersonal standpoint. All the other conditions for access to more specific principles would be empirical: knowledge of what practices or institutions would or would not serve the general welfare—leading to a long process of discovery, invention, trial, and error that can be expected to continue into the future. However, I do not accept this account because I believe the normative grounds of morality are not unitary but multiple: impartial benevolence is only one of them. Moral questions are posed and answered through the interaction and accommodation among a plurality of personal and impersonal values, some of which emerge only as the result of historical developments.

Even the universalizing step to the recognition that the interests of all humans, or all sentient creatures, have objective value and therefore provide anyone with reasons for action, starts a new set of questions. For if we accept as a contributing norm the impersonal value of the happiness or unhappiness of all sentient beings, that still doesn't tell us how these values are to be combined— when they conflict—to yield determinate reasons for action or policy. Impartial aggregative maximization is one possibility, but it is not the only one. If all we have to go on is the basic premise of universal value, which generates reasons to do what will promote happiness and prevent unhappiness, the most that strictly follows is that when reasons of this kind are unopposed—either for or against an action—they are decisive. When they conflict, they yield no determinate result by themselves, and further principles have to be brought to bear to determine what, if anything, is the right thing to do. Such further considerations are responsible for some of the variation in content of the actual working moralities that have been recognized historically and that are in operation today—including utilitarianism, which depends on substantive principles of impartiality, aggregation, and maximizing consequentialism. (Rawls famously argued that this answer makes the mistake of treating conflicts between the interests of different people like conflicts of interest within the life of a single person.)[16]

8.
—

Let me offer an interpretation of moral progress that is consistent with a pluralistic conception of the normative foundations of morality, together with the condition that moral requirements, however objective, apply only to beings that have the capacity to recognize them.

The need for these capacities is obvious in the cases of biological evolution and individual maturation—from hominid to human and from infant to adult. But I have been arguing, more controversially, that such developmental conditions for the application of moral norms extend also to cultural developments that are not biologically innate. Moral progress depends on the development of forms of thought and justification that provide access to moral reasons that were not accessible before. This is analogous to the distinction between the roles of basic reasoning capacities

---

16 John Rawls, *A Theory of Justice* (Harvard University Press, 1971), section 5.

and conceptual and empirical developments in determining what people had reason to believe at different stages in the history of scientific progress -- Twith the difference, as I have said, that in the moral case, the change in reasons constitutes a change in what is true, whereas in the scientific case it does not.

My alternative to the "expanding circle of impartiality" story is that morality, both individual and social or political, has evolved by continual interaction between personal and impersonal reasons, shaping institutions and practices that then generate new questions and new reasons. The process is one of repeated pursuit of reflective equilibrium, with each successive equilibrium being disrupted and transcended in its turn. I will elaborate by discussing briefly some examples of moral progress in four areas: (1) constitutional government and individual rights, (2) equality and comprehensive social justice, (3) sexual norms and social status, (4) global justice and the state.

To begin with the first topic: my understanding is opposed to the view that human rights are timeless and knowable in principle by rational reflection alone, and that the standard for evaluating the moral legitimacy of legal and political institutions is whether they protect or violate those pre-institutional rights. As I said above, the idea of a universal claim to the most basic liberty might have been in principle accessible even to those living in the world of ancient slavery, but the extensive and sophisticated system of individual rights that are now widely recognized in liberal democracies (though their precise content is contested) could not have been imagined, let alone acknowledged, except through a historical process.

This process was inseparable from the development of legal, political, and economic institutions and changing ideas about the conditions of their legitimacy. Earlier I discussed the example of freedom of expression and how the interpretation of that right is influenced by a certain conception of political legitimacy. But I believe that our understanding of all the major types of individual rights—freedom of religion and conscience, freedom of association, inviolability of the person, rights of privacy, sexual freedom—could only have evolved as part of a conception of the relation of the individual to the collective power of the state. That conception, of popular sovereignty, is now given institutional expression in liberal constitutional democracies, a relatively recent form of government. But these are not merely legal rights: to the extent that they are protected by law, it is because they have been recognized as rights that the state must secure to its members as a moral condition of its legitimacy.

Take freedom of religion, a distinctively modern concept. The Roman Empire tolerated a good deal of religious pluralism, but when it required sacrifices to the officially recognized deities, no one could have had the idea of objecting on the ground of a general, confessionally neutral right to freedom of worship. When Christians refused to make such sacrifices, it was because they believed the true religion, namely theirs, forbade it; and when in due course Christians came to command the power of the Roman state, they had no compunction about using it to suppress other religions. It was only after the horrible wars of religion following the Reformation and the decline of belief in a divine source of royal authority that the concept of a neutral right to individual freedom of religion became accessible to the inhabitants of Christendom. It could not have been understood apart from the idea of limits on sovereign authority that depend on an individualistic basis of that authority—its justifiability to each of those who are subjected to it. This therefore is an example of a moral principle becoming true only at a certain point in history, because the reasons for accepting it could not have been recognized earlier. Pre-modern Christians could have accepted some degree of toleration on purely practical grounds—because the suppression of heresy was too costly and disruptive of social peace—but not because it violated an individual right to freedom of worship.

The general point is that the delineation of these familiar human rights appears as the answer to a question that could not have been asked before a certain point in human history, namely: What limitations on the power of the state over the individual must be included in its basic constitution for those individuals to be able to regard its coercive authority over them as morally legitimate? It is a question that depends on the assumption that the state is granted its collective authority by its subjects, who have their own reasons for retaining certain spheres of autonomy for themselves. It cannot be answered by identifying a set of pre-political human rights and applying them as limitations on state power. For even if, implausibly, there were a pre-political "natural" right to freedom of expression, it could not on its own determine the answers to familiar questions about exceptions for libel, fraud, misleading commercial speech, hate speech, political campaign regulations, and so forth. The same holds for the detailed form of rights to privacy and for the elaborate systems of due process that govern limitations on individual liberty. These principles are

MORAL REALITY AND MORAL PROGRESS   59

subject to ongoing debate and refinement, but all of it is embedded in a concrete situation that poses specific questions about what should and should not be exempt from collective control or how the exercise of that control should be constrained.

The precursor of the complex system of individual rights is the rough sense that everyone is entitled to some sort of basic inviolability as part of their moral status. But the boundaries that define that inviolability cannot, I believe, be determined by pure reason. They depend on judgments about concrete situations that present us with different options. These judgments cannot be made at just any point in time by someone without the relevant experience. Even thought experiments would not help, because the thought experiments in which one is capable of exercising judgment depend on extrapolation from what one has experienced.

9.

To turn to the second topic: the accessibility of modern moral ideas about social and economic inequality depends on something beyond the conception of popular sovereignty that underlies the recognition of modern individual rights. It depends on a richer conception of the state—not as a mere framework of order and security making possible the peaceful coexistence and cooperation of its individual members, but as something much more consequential. It began with the recognition of a collective responsibility to provide protection against absolute destitution, through some form of public relief. This grew to include a guarantee of basic education, minimal health care, and regulation of working conditions. But the big step was recognition of the inescapable responsibility of the state, through its role in determining the structure of legal and economic institutions, for the distribution of social and economic advantages arising from the myriad transactions and activities that constitute the life of the society. This responsibility is both positive (what the state does) and negative (what it allows), and it is both causal and moral.

The moral outlook of democratic liberalism, which supports the modern welfare state, becomes accessible only as it emerges that the class structure of society can be modified by legal and institutional measures, short of revolution, that are compatible with the preservation of individual liberty. An essential change concerns the understanding of property rights. They have to be seen not merely as protected, but as created and defined by the state. This is done through laws governing incorporation and finance, taxation of all kinds and exemptions from taxation, wage regulation, and the provision (or not) of various forms of social security such as retirement, child benefits, unemployment and disability insurance, and medical care. The concept of distributive justice and the disputes over what it consists in get a foothold here, because members of a society come to see themselves as collectively responsible, through the state, for the huge effects that fiscal policies have on social and economic stratification. They have to ask themselves how this power should be exercised. The answers to such questions are notoriously controversial, but the questions cannot even be posed if one lacks this thick conception of state responsibility, which is available only in light of specific historical developments. Given that conception, we cannot avoid asking whether something should be done to limit the scope and effects of hereditary class inequalities. Even a policy of extreme laissez-faire becomes a positive choice, requiring a moral justification.

This is just one example of a type of question that the inhabitants of liberal democracies are constantly faced with: questions about the appropriate division of labor between the state and civil society. Such questions can be understood only from a point of view that permits moral assessment of the system as a whole, in light of a plurality of values that include individual liberty; collective responsibility; personal responsibility; fairness; security; and cultural, material, and communal flourishing. None of these factors can be captured by abstract definitions—their content is revealed only through reflection about concrete cases. Especially when they conflict, as they usually do, they can be realistically taken into account to yield a judgment only by someone historically situated, who has some grasp of the human and institutional circumstances that give rise to the question at hand. That is not just because without experience one cannot understand the relevant non-normative facts but because the relevant values themselves cannot be understood, weighed, and compared without the experience of their context and history of application.

10.

Morality has always had something to say about sex. But only recently have we been faced with the complex relation between norms of personal sexual conduct and norms of social justice pertaining to the inferior status of women compared to that of men, even in liberal societies without legally enforced subjugation. Of course, in much of the world the age-old disabilities and ferocious enforcement of unequal standards of conduct against women still prevail. These are the most important problems, demanding moral and social transformation in the face of implacable resistance from the men of those societies. But I want to discuss something subtler and more recent.

Catharine MacKinnon's argument that sexual harassment is a form of sex discrimination requires that we see it not merely as an interpersonal offense. It is that too, of course, and if it rises to the level of coercion—sex in exchange for continued employment, professional advancement, or a good grade—it is an abuse of institutional authority as well. But if it happens frequently—as appears to be the case—then the liability to harassment and the need to guard against it and deflect or otherwise deal with it becomes part of the social and institutional environment for many women, and part of the larger problem of the secondary status of women even in a society like ours with its ideals of legal equality. The offense of harassment is not merely an interpersonal transgression but a form of social injustice.

Note that this is true only of sexual harassment of women by men. When there is harassment of men by men, or women by women, or men by women—all of which certainly occur—these really are just interpersonal offenses or abuses of power. They may be very objectionable, but they do not reinforce or depend on any disadvantages of status for the sex to which their victims belong; they are not part of a pervasive problem of social justice. I believe that for this reason the institutional reactions to the cases of Kevin Spacey and James Levine were completely disproportionate—motivated by a punitive spirit toward sexual transgressions that has nothing to do with social justice.

We are now trying to find our feet in this complex terrain. The fact that sexual harassment is not merely an interpersonal offense doesn't mean that it's not a personal matter at all. And in defining its boundaries, we come up against the legitimate though contested claims of one of the most personal and important aspects of human life, and one that is particularly resistant to rational control. There are other cases, like the prohibition of racial discrimination in housing, where the old conservative objection that this was a matter of personal morality and not an appropriate subject for legislation now seems like a joke: the personal was rightly swallowed up by the political. But sex cannot and should not be so completely tamed. There is a current of exceptionally repressive opinion now at work which is trying to draw the boundaries much too strictly. Sexual attraction and its expression need not be disabling to its object, even if it is unwanted—provided the poor creep backs off after being rejected. But I would go further, and keep the lesser offenses of genuine sexual boorishness to be dealt with individually, by interpersonal deflection, exposure, and reproach.[17]

I believe we should adopt a principle of subsidiarity, instead of abolishing the public-private distinction completely. Misconduct in private life should not destroy professional or artistic status. Even in professional and institutional settings, up to a certain point,

---

17 We should also try to reverse the tendency to push the threshold of offense ever lower, so that it captures large swaths of mere insensitivity or cluelessness, like those that resulted in the railroading of Al Franken and Andrew Cuomo.

it should be possible to handle offensive behavior interpersonally and by effective norms of civility, rather than institutionally —unless of course a rejection or rebuke incurs institutional consequences or penalties. I realize that this is a hard line to draw, since the bruised male sexual ego may be inclined to retaliate. But a punitive policy toward every insensitive or clumsy manifestation of sexual interest becomes simply stultifying and is likely to inhibit many interactions that would be unobjectionable or even welcome. It ought to be acknowledged that sex is inherently likely to make men behave idiotically from time to time, and this in itself need not constitute an injury to the status of women. Sex is omnipresent and one of the best things in human life; it cannot flourish without some tolerance for its attendant embarrassments.

These are not easy questions. My main point is that they cannot be understood, and the disagreements about their correct answers cannot be pursued, except in the context of a historically specific set of values and circumstances that have brought them to light. These are, specifically, the radical liberation in the norms of sexual conduct that has followed on the availability of effective contraception, and the recognition that the unequal status of women is not just a matter of legal disabilities and discrimination but also of how they are regarded and treated informally. The clash between freedom and control in this case is very stark, but it could not be appreciated by someone who had not internalized these two conceptions.

This is just the latest step in a long process that began with a willingness to question the consequences of the difference between men's and women's roles in procreation for the social status of women. Instead of being regarded as natural, those consequences came gradually to be seen as requiring, and often lacking, moral justification—though it is only in the modern world, since the nineteenth century, that a truly egalitarian conception has been available. As with other cases of moral progress, it required that something that had seemed to be part of the natural order and taken as a given should come within the scope of human choice. It is true that Plato imagined a society whose ruling class would enjoy complete sexual equality, but this was a utopian conception that even he could not have imagined extending to society as a whole. And the questions we now face are still those of a society in which the sexes are systematically unequal. That is why personal interactions of a sexual nature raise questions of social justice.

11.

The question of what constitutes justice in the relations between sovereign states is very old, but it has taken a new form recently as a result of three developments. One is the spread of liberal egalitarian conceptions of intrastate social justice. Another is economic and informational globalization—a huge leap in interconnectedness. The third is the pressure for migration from poor regions of the world to rich ones, and the resistance it has evoked.

This is an example of moral progress only in the sense that new moral problems have been recognized: there is no sign of convergence on a set of right answers. But the problems, as in the other cases I have discussed, emerge only against a background of prior historical developments and cannot be understood apart from them. One question is whether the grounds for contemporary conceptions of socio-economic justice within a state (which of course include a substantial range of competing views) are such as to imply anything analogous at the international level. Even if intrastate justice is an associative obligation, dependent on our special relation to our fellow citizens, our relations to foreign populations are growing institutionally and economically thicker all the time, and this may generate new forms of associative obligation.

But the globalization of justice collides with the robust conviction that each nation has the right to give priority to the interests of its own citizens. It is generally conceded that this right is not unlimited: wars of conquest are now widely regarded as unacceptable (notwithstanding Russia's current defiance of that norm in Ukraine), and many nations recognize an obligation to give asylum to political refugees. But it is an open question how the claims of national interest should respond to increasing pressure from global emergencies. And what about the conditions of membership itself? Immigration, legal and illegal, to Europe and North America poses acute and familiar problems: whether the interest of economic refugees in escaping desperate conditions in ill-governed states should carry any weight in the policies of rich states; whether the danger that a large and unassimilable immigrant population will erode a valued social and political culture is a ground for selective exclusion.

Over the course of the twentieth century, our civilization has converged on a conception of social justice that recognizes a collective responsibility of the members of a society toward one another. It is the responsibility to ensure that the powers of the state, conceived as a common enterprise, are employed to create a structure for their social and economic activities—an impersonal system of

property, fiscality, and provision of public goods—that instanti-
ates a form of social justice for all, even if significant inequalities
remain. The welfare state, more or less. But because the obligation
is conceived as associative, it depends on our common membership
in the same society. We do not have this collective responsibility
with regard to foreigners; and we have no obligation to admit outsiders merely because they would like to benefit, by association,
from our collective obligation. The inclusivity of this conception
of social justice is also exclusive.

Today this conception of justice is under enormous pressure
from the growth of global interconnections and the mobility of
populations. International inequalities seem to pose a question of
fundamental justice, to which we have no convincing response. It
is the following question: Can we find principles of international
or global governance and conduct that are morally and institutionally compatible with our conception of state or domestic
justice? To answer that question is both a philosophical and an
institutional task, like the earlier task of reconciling the rights of
private property with the values of social justice. But the answer
remains to be discovered, and I believe it will require a form of
moral progress that is not just the recognition of what has been
true all along, but an expansion of morality.

The examples I have discussed all take the form of questions
for people who find themselves in a particular historical situation.
Those who attempt to answer them must rely on moral norms
that have come to be recognized over time but that may have to
be changed or supplemented to deal with new circumstances.
Perhaps in some cases such answers could have been discovered
long in advance, by imaginative thought experiments about future

possibilities. But I believe this will often not be the case, particularly when the imagined future includes normative developments that are essential in posing the question. Someone living in the sixteenth century could not by sheer imagination have had access to the liberal democratic conception of social justice that poses the present dilemma of global versus national obligations. The question of what ought to be done always starts from the norms one accepts now, even if in the course of trying to answer it one concludes that they may have to be changed.

12.

The gradual progress toward modern moral conceptions has been path-dependent—dependent on the development of new institutions or new forms of thought, usually in combination. I have argued that these advances often provide examples of how the accessibility of specific moral reasons can depend on something more than a general faculty of practical reason—something that is available only as a result of contingent historical developments. So far I have discussed this phenomenon only with respect to the relation between the present and the past. But it has obvious implications for the relation between the present and the future. We can expect the process to continue—for moral progress to occur which consists partly in the emergence of new moral truths that are not accessible to us, as we now are. These will depend on conceptual and normative and institutional developments that can be effectively grasped, for purposes of moral reflection, only by people who actually live through them. They are not now accessible by pure reason, even with the aid of an adventurous science-fictional imagination.

But I don't want to let us off the hook completely. There will certainly also be moral progress that consists in the discovery of truths that we are now too pig-headed, dishonest, or self-deceiving to recognize, but that are nevertheless accessible to us and therefore already apply to us. Both kinds of progress are to be expected, and we should regard our present moral convictions with a certain humility in view of the contingency of our place in history. (We should have the same attitude toward our present stage of scientific understanding, though in both areas it is remarkably difficult for human beings not to think they are approaching the finish line.)

I expect that future developments will continue to be dominated by the attempt to find an accommodation among reasons arising from three familiar perspectives: the personal, the associative or social, and the impersonal. Progress will require changes to existing institutions, creation of new institutions, and consonant changes in norms for both institutions and individuals. For example, as noted in the previous section, there is clearly a need for greater world governance, but we have no non-utopian conception of the conditions of legitimacy for a strong international authority. In view of the plain facts about our world, global legitimacy can't be designed by analogy with the conditions of democratic legitimacy for the nation-state. There is an upper bound to the divergence of interests, values, language, and culture that can be contained within a legitimate democratic polity. Yet eventually a new structure may emerge, through contingent historical developments, which will be governed by norms that could not have been recognized in its absence but that become accessible to people who live in the world so structured.

Another area in which the future may see large changes is in the norms and institutions governing sexual relations, sexual conduct, procreation, and familial obligations. There are at present huge differences between cultures about these matters, and within Western societies the situation is very fluid, confusing, and conflict-ridden. Whether a stable accommodation that respects individual autonomy, equality of status, and collective welfare can emerge from the present sexual tangle is far from clear, but some development is clearly needed.

This is to project into the future a form of moral progress similar to what has happened in the past. Perhaps I am being too conservative, and humanity will develop much more radically. Certainly there are those who think it should. On the one hand, there is Nietzsche's call to free ourselves of the life-destroying tyranny of the impersonal point of view. On the other hand, there is the utilitarian project of establishing the hegemony of the impersonal point of view by requiring all deontological standards such as individual rights or associative obligations to pass a consequentialist test or else be revised or abandoned. I can speak only from where I am, but it seems to me that either of those revolutions, if they occurred, would not be moral progress but a loss of moral knowledge. As for other moral revolutions, unimaginable by us, I can say only that I suspect they will happen.

INDEX

For the benefit of digital users, indexed terms that span two pages (e.g., 52–53) may, on occasion, appear on only one of those pages.